HAWAII Tropical Botanical GARDEN

A Garden in a Valley on the Ocean

Hawaii Tropical Botanical Garden
P.O. Box 80, Papaikou, Hawaii 96781 Telephone 808/964-5233
Website: www.hawaiigarden.com Email: htbg@ilhawaii.net

Produced by Editorial Services, PO Box 3155, Santa Barbara, CA 93130
Content Advisors: Dan and Pauline Lutkenhouse
Taxonomist: Sean Callahan
Project Manager/Writer: Cynthia Anderson
Designer: Britta Bonette
Copy editor: Courtney C. Fischer
Text research: Mark J. Rauzon

Printed in Hong Kong by WMG HKG LTD.
ISBN 0-9639711-0-7 (softcover) ISBN 0-9639711-1-5 (hardcover)

Photo Credits
Richard Cook III: page 59. **Amy Evans**: page 9 bottom inset, 21 bottom, 24 bottom, 25, 29 top, 32 top inset and lower left, 34 right inset, 41 right, 42 right, 47 center inset, 48, 49, 53 bottom, 57, 60, 61 bottom, 67 bottom, 69 inset, 70 left, 74 top left, 75 top, 76 left, 78 top and bottom right, 83 center left, 85 bottom, 87 background, 88 right, 95. **Jack Jeffrey:** page 68 center and bottom, 88 left, back flap top. **Ken Komata:** page 9 top. **Light, Inc.:** pages 18-19, 20 top, 26-27, 68 top. **Dan J. Lutkenhouse:** title page, copyright page, dedication page center, contents left and right top, page 9 bottom, 12 left, 13 bottom, 15 top, 16 left, 28, 33, 34 left, 35-39, 43-45, 47 top left, 51 right, 54, 56 bottom, 58 inset, 61 top, 63, 64 bottom, 65, 67 top left, 70 right, 72, 75 inset, 77, 79 bottom and inset, 80-81, 82 lower right, 83 top, center right, bottom, 85 top, 92, back cover left. **Pauline Lutkenhouse:** page 15 bottom, 16 bottom right. **Lyman Museum:** page 20 bottom, 21 top, 22, 23. **Mark J. Rauzon:** front cover inset, front flap, inside front cover, dedication page top, contents page right center (Parts 1 & 2), page 10, 11, 13 top, 16 top right, 24 top, 29 bottom, 30, 31, 32 top left, 34 top center, 40, 42 left, 46, 47 top right and bottom, 50, 51 bottom left, 52, 53 top, 55, 56 top, 62 left, 64 top, 66, 67 top right, 69 top left and right, 71, 73 bottom, 74 bottom, 76 top and bottom right, 79 top right, 82 top left and right, 84, 86 top, 87 inset, 89, 90, 93, 94, back flap bottom, back cover center and right. **Reed and Jay Photographers:** page 8. **Greg Vaughn:** front cover waterfall, dedication page bottom, contents page bottom right, page 14, 73 right, 86 bottom, inside back cover.

Front cover photo: Onomea Falls. Inset: Hawaiian pink hibiscus *(Hibiscus hybrid)*. Front flap: Onomea Bay. Back flap: top, Red liwibird; bottom, Pink plumeria *(Plumeria rubra)*. Back cover photos: Tropical water lily *(Nymphaea var.)*, Hanging lobster claw heliconia *(Heliconia rostrata)*, and orange plumeria *(Plumeria rubra)*. This page: Queens crepe myrtle *(Lagerstroemia)*. Opposite page: Pink hibiscus *(Hibiscus hybrid)*; Passion fruit flower *(Passiflora quadrangularis)*.

DEDICATION

This Garden Book is dedicated to my dear, sweet Mother, Ethelind M. Everts Lutkenhouse, who taught me at a very young age to love and appreciate the beauty of nature — to plant a seed each day and carefully watch it grow into a beautiful flower or tree. In addition, in remembrance of my Father, John F. Lutkenhouse, a civil engineer, who taught me to build and create, to work hard and strive to leave this world a better place for mankind. Last, but not least, to my Wife, Pauline, whose patience and understanding allowed me to work in the jungle for more than seventeen years to create a place of beauty.

Dan J. Lutkenhouse,
Founder

CONTENTS

PROLOGUE

PART 1

PART 2

PART 3

FOUNDERS OF THE GARDEN, DAN AND PAULINE LUTKENHOUSE AND THEIR DOG, FAWN.

The Hawaiian word "Aloha" means Love. Hawaii Tropical Botanical Garden is the creation of my husband, Dan. Everywhere one walks within the Garden, you feel the "Aloha" of his creation — in the BEAUTY of the plants and trees and flowers selected by him — in the GENTLENESS of the winding trails designed by him — in the SERENITY of the ocean and waterfalls which were brought right to your footsteps — in the EXCITEMENT of the colors, vibrant flowers lovingly displayed to awaken one's senses — and in the STRENGTH of the tall trees which he nurtured and saved to make one feel protected and safe.

This describes the Garden. This describes the man's artistry — it also describes his qualities.

And for me, it portrays our life together — a life of "Aloha" and one that I wouldn't have missed for the world!

Pauline Lutkenhouse

ABOVE: ROSE GRAPE (*MEDINILLA MAGNIFICA*).
LEFT: HAWAII TROPICAL BOTANICAL GARDEN'S MAGNIFICENT OCEANFRONT SETTING.

PROLOGUE

He that plants trees
loves others besides himself.
— English proverb

Aloha and welcome to Onomea Bay and the Hawaii Tropical Botanical Garden! This "garden in a valley on the ocean" is located on the lush Hamakua Coast, seven miles northeast of Hilo in the sheltered Onomea Valley. In Hawaii, *onomea* means "the best place". And indeed, no better place exists.

The Hawaii Tropical Botanical Garden is a museum of living plants that attracts photographers, gardeners, botanists, scientists, and nature lovers from around the world. The Garden's collection of tropical plants is international in scope. Over 2,000 species, representing more than 125 families and 750 genera, are found in this one-of-a-kind garden. The 40-acre valley is a natural greenhouse, protected from buffeting tradewinds and blessed with fertile volcanic soil. Some of the Garden's enormous mango and coconut palm trees are over 100 years old. Tropical plants that struggle to grow in homes and gardens across America reach gigantic proportions here.

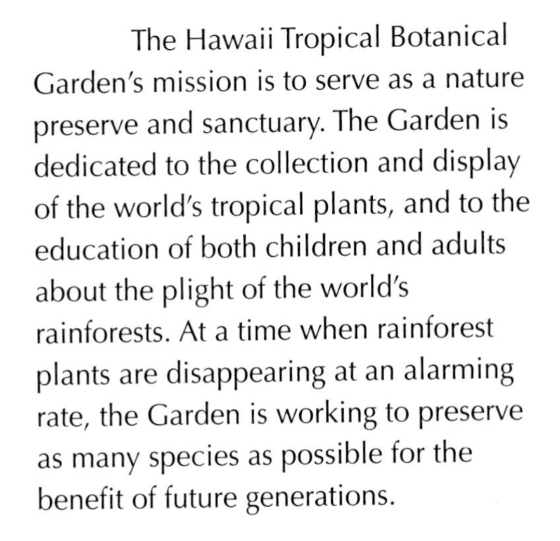

The Hawaii Tropical Botanical Garden's mission is to serve as a nature preserve and sanctuary. The Garden is dedicated to the collection and display of the world's tropical plants, and to the education of both children and adults about the plight of the world's rainforests. At a time when rainforest plants are disappearing at an alarming rate, the Garden is working to preserve as many species as possible for the benefit of future generations.

WHITE-FLOWERING CAT'S WHISKERS (*ORTHOSIPHON STAMINEUS*).

The Vision of Dan Lutkenhouse

The Garden was created through the untiring efforts of one man, Dan J. Lutkenhouse, who discovered Onomea Valley in 1977 while vacationing with his wife, Pauline. Lutkenhouse purchased the 17-acre parcel for its seclusion and beauty, without knowing exactly what to do with it. Quickly abandoning ideas for a commercial venture which would destroy the natural environment, he decided instead to establish a botanical garden to preserve the valley and its beauty forever.

When first located by the Lutkenhouses, Onomea Valley was an overgrown and virtually impenetrable jungle, choked with wild invasive trees, weed and thorn thickets, and strangling vines. Lutkenhouse sold his 40-year-old trucking business in San Francisco and moved to the island of Hawaii in order to devote himself full time to the development of the Garden.

Every day for eight years, Pauline would pack Dan a brown bag lunch and he would disappear into the jungle, returning at night dirty and tired, but happy, During that time Dan, his assistant Terry Takiue, and two helpers worked with cane knives, sickles, picks,

ABOVE: THE FLOWERING BEAUTY OF THE AFRICAN TULIP TREE (*SPATHODEA CAMPANULATA*), IN THE DENSE GARDEN JUNGLE. FAR RIGHT: A TRANQUIL VIEW OF LILY LAKE.

shovels, and a chain saw clearing paths through the jungle. All the work was done by hand to avoid disturbing the natural environment or destroying valuable plants and tree roots. The men kept a slow and easy pace, so as not to suffer heat stroke or dehydration in the steamy jungle. The work would continue seven days a week until the Garden opened to the public in 1984.

Trails were hewn from hard lava rock with picks and shovels. To keep the soil from compacting and the

natural beauty from being destroyed, no tractors were used; excess rock was removed and gravel brought in by wheelbarrow. Lutkenhouse followed the contours of the land in designing the Garden trails, which curve and wind their way throughout the jungle. Gradually, secret landscapes revealed themselves. It took years of carefully clearing the jungle before Lutkenhouse discovered the crown jewel of the Garden—a three-tiered waterfall said to be the most beautiful in all Hawaii.

Though Lutkenhouse has no formal botanical training, with his love of nature he has created a living tapestry in keeping with the intimate nature of the site. Subtle vistas unfold as you meander along the Garden paths. Patterned foliage and brilliantly colored flowers invite close inspection, enticing you further into the mysteries of the jungle. This is the allure of paradise. The Hawaiians have a word for it—*aina*, or "the spirit of the land."

Lutkenhouse himself has chosen the location of every plant and tree introduced to the Garden. From the Lily Lake Vista, more species of plants can be seen in one place than anywhere else on earth. Over 110 species have been counted within this viewshed, most planted by Lutkenhouse and his staff. This vivid experience of the tropics has been enriched by the plant collecting trips Dan and Pauline Lutkenhouse have taken to tropical jungles around the world.

A Tropical Adventure

Africa, Australia, Bali,Indonesia, and Madagascar are just a few of the exotic destinations the Lutkenhouses have visited in search of rare and endangered tropical plants. Their trip to Madagascar, which is located off the coast of Africa, was among the most memorable. They visited remote areas, stayed in native villages, and viewed rare plants that few Americans have ever seen. They found the local residents to be kind, shy, gentle, and

THIS PAGE: TWO VIEWS OF THE MADAGASCAR VILLAGE WHERE THE LUTKENHOUSES OBTAINED A RARE STONE PLANT (*ODOSICYOS SP.*).
OPPOSITE: FOUNDER OF THE GARDEN DAN LUTKENHOUSE IS DWARFED BY A GIANT RED LOBSTER CLAW HELICONIA (*HELICONIA CARIBAEA C.V. PURPUREA*), ONE OF THE WORLD'S LARGEST SPECIMENS.

peaceful people living in extremely primitive conditions. After weeks of bumping along gutted dirt roads in a Land Rover with temperatures exceeding 100 degrees, they came to a village where they saw an extremely rare Madagascar stone plant lying next to a native grass hut.

After an hour of negotiation with the plant's finder, the chief, and an interpreter, the Lutkenhouses were able to buy this rare and unusual plant for $3. For the long trip back to Hawaii, Lutkenhouse hand-carried the plant in a jute bag. En route the plant started to grow, much to his amazement and the bewilderment of customs officials on three continents who had never seen such a plant.

The base of the Madagascar stone plant resembles a crude rock bowl approximately two feet across. Each year it produces a long green vine from the center of the woody plant, and a yellow flower. Little is known about this rare plant; the genus is *Odosicyos*, but the species is unknown. It is believed to be related to the cucumber family.

The Garden is now home to many species of plants from Madagascar. In the future Dan Lutkenhouse hopes to obtain more plants from this unique island.

BELOW: PAULINE LUTKENHOUSE STANDS BY A BAOBAB TREE (*ADANSONIA DIGITATA*) IN MADAGASCAR.

Preserving the Spirit of the Land

To protect the Garden site, Dan and Pauline Lutkenhouse have established a non-profit 501(c)(3) corporation and have taken legal steps to insure the land will never be sold or commercially developed. Dan is adamant on this point. "It's too precious a valley to be developed. We're preserving the valley so that mankind can enjoy it forever."

TOP: DAN LUTKENHOUSE AND THE MADAGASCAR STONE PLANT (*ODOSICYOS SP.*). ABOVE: DAN WITH A SPECIMEN OF PACHYPODIUM (*PACHYPODIUM SP.*).

He adds, "I believe that we should all try to leave the world a better place than we found it."

From a diamond in the rough, the Hawaii Tropical Botanical Garden is being polished to perfection. While the star of the show is the Garden itself, its creation and success are attributed to more than 17 years of hard work and dedication, as well as the prior business experience of Dan

and Pauline. Today the Garden has 17 full-time employees and is financially self-supporting. Dan and Pauline gain no financial rewards from the Garden; instead, they have contributed more than $2 million of their own personal funds to establish it. Their reward is the true enjoyment the Garden provides to its visitors. Dan and Pauline's vision of preserving rare tropical plants in one of Hawaii's most beautiful natural settings has been shared by more than 500,000 visitors to date.

The Lutkenhouses have listened to the land and its creator, and allowed the *aina* to guide them. Under their protection, the spectacular flora and fauna of Onomea Bay are flourishing as a world-class botanical garden, described by many as the most beautiful accessible tropical jungle garden in the world. Truly the Garden is a perfect expression of the state motto of Hawaii, *ua mau ke ea o ka aina i ka pono*, or "the life of the land is perpetuated in righteousness."

TOP: VISITORS ENJOY THE GARDEN'S WINDING TRAILS. ABOVE: AFRICAN TULIP TREE BLOSSOM (*SPATHODEA CAMPANULATA*). RIGHT: VIEW ALONG COOK PINE TRAIL.

PART 1

Onomea Bay: A Geological and Human History

E ola mau, e Pele e!
'Eli'eli kau mai!
Long life to you, Pele!
—Hawaiian Goddess of Volcanoes

The Hawaiian Islands are actually a chain of volcanic mountains, described by Mark Twain as "the loveliest fleet of islands to sail in any ocean." Twain was more than flattering in his analogy; he was also scientifically correct. The Hawaiian Islands are slowly moving through the Pacific Ocean in a northwesterly direction, the result of a phenomenon known as plate tectonics.

Geologists have determined that the Hawaiian Islands lie on the upper crust of the Pacific continental plate. The plate is the size of half the North Pacific Ocean, and it literally floats on the heavier magma of the earth's core. A crack in the continental plate leaks lava to the upper surface. This is the "hot spot" that has created the entire Hawaiian archipelago. Lava pours out of the ocean floor and piles up as hardened magma until the newly formed land reaches above sea level.

RIGHT: THE BIG ISLAND IS A HOTBED OF VOLCANIC ACTIVITIY. BELOW: ONOMEA ARCH IN THE LATE NINETEENTH CENTURY. PRECEDING PAGE: FOR OVER 11 YEARS, MOLTEN LAVA HAS FLOWED DOWN TO THE OCEAN FROM A VENT CALLED *PUU-OO* ON KILAUEA MOUNTAIN — ABOUT 50 MILES FROM THE GARDEN.

Today "hot spots" still smoke on the Big Island of Hawaii, the youngest island in the chain, just as they did millions of years ago. The Kilauea Volcano in Hawaii Volcanoes National Park is very active; molten lava flows down to the ocean, creating more new land on the island of Hawaii. The next Hawaiian island, named Loihi, is a mountain of fresh lava slowly growing on the sea floor southeast of the Big Island. In several thousand years it will rise above the ocean surface, and another Hawaiian island will be added to the chain.

The Shaping of Onomea Bay

Over the millennia, erosion has been one of the primary forces shaping Onomea Bay. Onomea and Alakahi streams have carved the valley, while winds and waves have cut the lava cliffs. Earthquakes and tsunamis have rocked the coast, causing radical changes in the face of the landscape.

The most notable work of the elements was Onomea Arch, carved from the cliffs by restless waterpower. Legend has it that King Kamehameha threw his spear to create this huge tunnel in the rock. A famous landmark, the arch attracted visitors to Onomea Bay long before the Garden was established.

Onomea Arch fell during an earthquake in 1956 after standing for thousands of years. Today the fallen arch appears as a wide crevice in the cliff on the north side of Onomea Bay, but this favorite Hilo landmark is preserved in antique postcards which recall its glory from the turn of the century.

ABOVE: AN INTER-ISLAND SAILING SHIP USED TO TRANSPORT SUGAR AND SUPPLIES.
RIGHT: ONE OF SEVERAL OLD GRAVESITES IN THE GARDEN VALLEY. EXACTLY WHO RESTS HERE IS UNKNOWN.

The Bay's Early History

Long ago, Onomea Bay was a fishing village for the early Hawaiians. Old stone walls in the Garden today were created by early settlers to make terraces for growing taro and sugar cane. These stone walls kept the land on the slope from eroding into the stream.

Onomea Bay served as one of the Big Island's first natural landing areas for sailing ships. In the early 1800s the fishing village, known as

ABOVE: TWO VIEWS OF OLD ONOMEA VALLEY, C. 1850. SOME VEGETATION WAS REMOVED BY NATIVE HAWAIIANS; MORE EXTENSIVE CLEARING WAS DONE BY EARLY SETTLERS.

Kahali'i, became a shipping port, first importing materials to construct the Onomea Sugar Mill and then exporting raw sugar. The settlers were a mixture of Portuguese, Chinese, Japanese, and Filipinos who came here to work in the sugar cane fields and build the Onomea Sugar Mill.

In the hills above the Garden are relics of the Onomea Sugar Mill. Rusty iron trestles and flumes once stood where hand-cut cane was floated to the mill. Sacks of unrefined sugar were then loaded on donkeys and taken down a trail to the docks.

Another remnant of this era is a Portuguese bake oven, discovered on

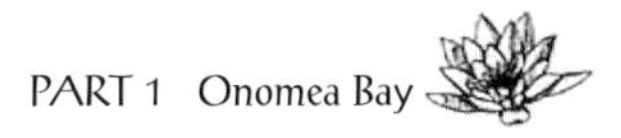

ABOVE: HAWAIIAN GRASS HUTS.
LEFT: THE OLD CHURCH ON THE SCENIC ROUTE, FORMERLY THE GARDEN VISITORS CENTER, WHICH BURNED IN 1988.

ABOVE: AN OLD PORTUGUESE BAKE OVEN, C. 1850, DISCOVERED WHILE CLEARING THE JUNGLE. RIGHT: HANGING BANANAS *(MUSA PARADISIACA)*.

Cook Pine Trail when the jungle was being cleared. Made of heat resistant rocks, the igloo-like structure was used to bake bread on a flat shovel.

After the Onomea Sugar Mill ceased operations, the early settlers gradually moved away. Later, part of the valley was a *lilikoi*, or passionfruit farm. Some cattle were grazed in the valley as well. A row of stately old palms that lines the narrow scenic route was planted by the plantation supervisor from the Onomea Sugar Mill. Wild banana, mango, coconut, and guava trees planted by the early settlers have reached towering heights and still grow here today.

Onomea Valley was deserted in the early 1900s, and the vegetation grew so densely that few signs of former habitation could be seen. The last resident of the valley was Lono Waikii. Legend has it that Lono's wife would get angry at his all-day fishing and drinking adventures and hide his whiskey bottles in a banyan tree. The tree grew up around the bottles, where they can still be seen today.

In Harmony with Nature

As can be seen in historical photos of Onomea Valley, the early settlers removed all of the valley's native vegetation. There remained only some tall coconut palms, which now are over 150 years old. The tall mango and monkeypod trees in the valley today have grown up since 1850.

Dan and Pauline Lutkenhouse have transformed Onomea Valley from a dense jungle to a pristine tropical paradise. Plant specimens have been gathered from tropical jungles around the world and planted by hand. The entire valley is treated as a nature

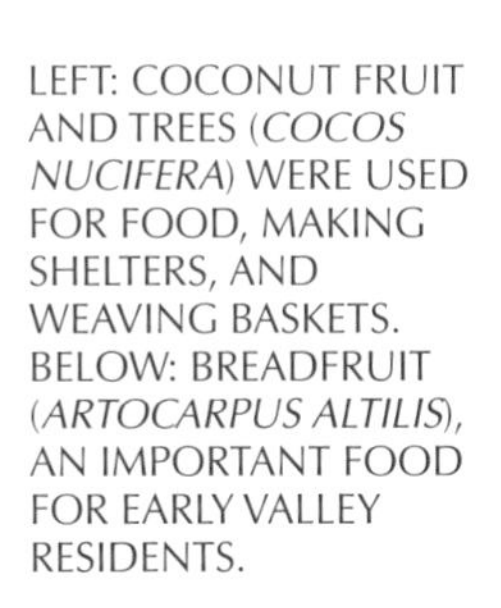

LEFT: COCONUT FRUIT AND TREES (*COCOS NUCIFERA*) WERE USED FOR FOOD, MAKING SHELTERS, AND WEAVING BASKETS. BELOW: BREADFRUIT (*ARTOCARPUS ALTILIS*), AN IMPORTANT FOOD FOR EARLY VALLEY RESIDENTS.

preserve. To protect the environment, no cars or tour buses are allowed in the valley; visitors park a half-mile from the Garden entrance and are transported down into the Garden by mini-bus.

Visitors marvel at the giant rainforest trees that form the dense overstory canopy of Onomea Valley. These include 100-foot tall mango trees that were probably planted in the late 1800s; breadfruit trees that date from the early 1900s, with their huge, speckled green fruit; and towering coconut palms that are constantly pruned to protect visitors from falling fruit. Low-growing taro plants can also be seen, carryovers from village life at the turn of the century when starchy poi, made from taro roots, was a staple of the Hawaiian diet.

In the Hawaii Tropical Botanical Garden, exotic plants gathered from distant tropical rainforests grow side by side with native Hawaiian plants. Together they form a spectacular living work of art in the only tropical botanical garden in the United States that is situated on an ocean coast.

SEA TURTLES AND BRIGHTLY COLORED FISH THRIVE IN THE WATERS OF ONOMEA BAY.

Marine Life

Along the shores of Onomea Bay, the patient visitor can observe abundant marine life — everything from single-celled yellow algae to sea turtles.

Creatures of the intertidal community include *opihi,* mollusks that look like chinamen's hats; and *purplevana,* or spineless sea urchins. *A'ama* crabs prance on pointly claws and cling to the wave-washed lava rocks. Freshwater prawns swim in Onomea and Alakahi streams, while farther out to sea, spiny lobsters and giant sea turtles make their home.

Onomea Bay is visited by two species of endangered sea turtles: the green sea turtle, or *honu,* and the hawksbill sea turtle, or *'ea.* In March 1992, a monk seal was found sleeping on the Garden's beach at Crab Cove — the first time this rare and endangered mammal has been sighted on the Big Island. The most ancient of all living seals, the monk seal is endemic to Hawaii, yet today rarely visits the inhabited main islands. Its presence at Onomea Bay is an important sign of the Garden's success in creating a peaceful marine sanctuary.

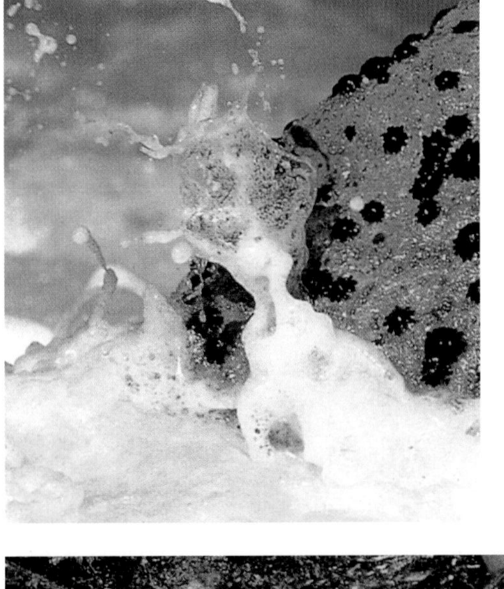

SEA URCHINS AND *A'AMA* CRABS DWELL IN THE INTERTIDAL ZONE.

PART 2

A Day in Hawaii Tropical Botanical Garden

The kiss of the sun for pardon,
The song of the bird for mirth,
One is nearer God's heart in a garden
Than anywhere else on earth.
—Author unknown

TOP: THE GARDEN'S CUSTOM DESIGNED "FLIP FLOP GATE," AT THE ENTRANCE TO THE JEEP TRAIL.
ABOVE: BRIGHT RED BOUGAINVILLEA (*BOUGAINVILLEA HYBRID*).
OPPOSITE: A VIEW OF THE GARDEN JUNGLE.

The day begins at the Garden's Visitor Center and Gift Shop, located seven miles north of Hilo on the four-mile Scenic Route near Papaikou. After a brief introduction, visitors board one of the Garden's mini-buses for a five-minute ride along the Scenic Route and then down into the ocean valley.

The mini-bus stops at an overlook above Turtle Bay and Crab Cove. From here lucky visitors might spy a sea turtle surfacing for a breath of air. The bay is usually peaceful, but sometimes huge waves born thousands of miles to the south come crashing onto the spectacular rocky shoreline.

A one-of-a-kind gate, designed by Dan Lutkenhouse, marks the entrance to the Garden. The gate is constructed of iron panels about three feet high, which are secured to a giant hinge and attached to iron weights. The vehicle's front tires push the gate open and the iron weights then pull it closed

ABOVE: SMALL BROMELIAD CIRCLE. BROMELIAD SPECIES CAN BE SEEN THROUGHOUT THE GARDEN.

A MINI-BUS TRANSPORTS VISITORS TO AND FROM THE GARDEN.

automatically — all without electricity. Necessity is the mother of this invention, since there is no electricity in the Garden.

The only entrance to the Garden in a Valley is this one-way narrow road, known as the "Jeep Trail." As the trail borders on a steep cliff, the descent into the Garden is exciting, with spectacular ocean views. Down in the valley, about every 20 minutes, the mini-bus drops off and picks up passengers at the turn-around by a group of immense mango trees.

Rain Shelter

Under the deep shade of the mango trees is the rain shelter, the starting point for self-guided walking tours of the Garden. Here Garden staff members give informal talks and answer questions. A generous supply of umbrellas and mosquito repellent is available. Each visitor is provided with a trail map for the self-guided tour along the rainforest trails.

The Hilo side of the Big Island is blessed with frequent rain, with over 160 inches falling annually. Rain brings out the best in the Garden. The many hues in green vegetation sparkle and gleam, and the moisture unlocks hidden fragrances of blooms and bark. Jungle plants thrive, some growing as much as six inches in one day.

Trail 1 to the Ocean

From the rain shelter, the first trail leads to the majestic Pacific coast, passing giant coconut palms and monkeypod trees on the way. Some of these trees are more than 100 years

TOP: SELF-GUIDED TOURS BEGIN AT THE RAIN SHELTER.
RIGHT: WALKING THE GARDEN TRAILS IS EASY AND SAFE.
BELOW: RARE BAT PLANT (*TACCA CHANTRIERI*).

TOP RIGHT: *CALATHEA WARSCEWICZII.*
ABOVE: PAULINE LUTKENHOUSE ALONG THE TRAIL.
LEFT: PINK CATTLEYA ORCHIDS (*CATTLEYA HYBRID*).
OPPOSITE: LAVA TUBE CAVE ALONG THE OCEANFRONT TRAIL.

old, and as ancestors from the past, they have witnessed the migration of settlers in and out of the valley. Australian tree ferns provide a lacy, vivid green backdrop. Graceful licuala palms are found here as well.

More than 60 different species are found on this trail. (See plant list, page 88.) Exotic plants include Indonesian ginger, Amazon lily, Philippine orchids, and beautiful bromeliads. Look for the rare bat plant, whose black flowers open only for a day. The light playing over the rainforest canopy provides an ever-changing backdrop that contrasts with patterns of leaves along the trail.

The Oceanfront Trail

Along the Oceanfront Trail, ironwood trees from Australia whisper in the tradewinds. Majestic breadfruit trees are plentiful; these provided a staple food for the ancient Hawaiians and early settlers, and chewing gum was made from its white, milky sap.

ABOVE, LEFT AND RIGHT: BROMELIADS (*VRIESEA SP.*). ABOVE CENTER: CROTON (*CODIAEUM VARIEGATUM*). OPPOSITE PAGE: VIEWS FROM THE OCEANFRONT TRAIL. NEXT PAGE: OCEANFRONT VISTA POINT.

Lauhala trees, which the early Hawaiians used to make tapa mats, are set back from the short cliff overlooking Onomea Bay. The atmosphere is timeless, with the rolling breakers inspiring peaceful meditation.

Nearby in a secluded setting are four unmarked graves of persons who once lived in the valley. Below on the coast is the lava tube of Onomea Bay.

Lava tubes are formed by hot lava flowing from vents into the sea. The outer layer of lava cools into rock, insulating the still molten interior through which the lava pours on its way to the sea. Today, the tube is filled with sea water and beaten by breaking waves. No one knows how far the lava tube runs into the valley. This is a very interesting and beautiful feature of the rugged shoreline of the Garden.

Presently there are at least 28 different species of plants growing on the Oceanfront Trail. (See plant list, page 88).

The Legend of Twin Rocks

The village of Kahali'i was located on a large point of land which extends into Onomea Bay.Though the village is gone,the descendants of Kahali'i residents still remember some of the legends concerning the area's landmarks. One story tells of the origin of two rock formations at the head of Onomea Bay that are said to be a young man and woman, known as the lovers of Kahali'i.

Legend has it that one day, a chief of the village spotted many canoes with sails heading shoreward in their direction. Fearing an attack, the chiefs and family elders held a council to determine a course of action. They decided to build a reef to prevent a landing on their beaches. Not having the means to complete the task quickly enough, they asked that a young man and a young woman be the guides and protectors of the village by giving their lives. Two willing individuals were found.

That night a decree was sent to all who lived at Kahali'i to remain indoors from sunset to sunrise without making any light or sound, on penalty of death. During the darkest hour of the night, steps were heard walking through the village, and then silence prevailed until morning. In the light of the new day, the people went down to the shoreline where they were amazed to find the lovers gone, and in their place two gigantic rock formations at the entrance to the bay, along with many other smaller rocks strewn about, as if on guard.

The chief informed the people that no canoe could pass the treacherous currents swirling around the rocks unless allowed to do so by the guardians. The lovers and their offspring still stand today, sentinels at the head of the bay.

The Naupaka Flower

Growing on the cliffs along the bayfront are leafy naupaka shrubs with small white half-flowers. According to legend, the volcano goddess Pele, disguised as a beautiful woman, fell in love with a village youth. When he rejected her and returned to his sweetheart, she tore him away and pursued him into the mountains. There the gods took pity on him and turned him into a half-flower, the mountain naupaka. To escape Pele's wrath, his sweetheart was turned into a matching half-flower, the beach naupaka. The lovers bloom forever, eternally separated.

LEFT: CRAB COVE BEACH.
RIGHT: THE LEGENDARY BEACH NAUPAKA FLOWER *(SCAEVOLA FRUTESCENS).*

Rock Island and Crab Cove

As important residents of the Garden's marine preserve, *a'ama* crabs and *opihi* thrive here. Onomea Bay is one of the few places on the Big Island where these marine creatures are not hunted. From Turtle Point, you can look down over Crab Cove; further along the shore, at Turtle Bay Vista, you can see the black sand beach where a monk seal recently made an unprecedented visit.

Wooden benches invite rest, reflection, and contemplation above the rocky shore. This spot is one of the most exhilarating and restful in all the Garden.

From Crab Cove, visitors follow the Alakahi Stream through a palm forest, returning to Lily Lake and other beautiful Garden trails.

Alakahi ("Litter River") Stream Trail

This bubbling stream is crossed by two small bridges, made like all of the Garden's bridges from wood and metal salvaged from old sugar mills. The stream is surrounded by orange heliconias, giant mango trees, and pink ornamental bananas. Approaching Lily Lake, an abundance of anthuriums are present, including obakes, a beautiful multicolored Japanese hybrid variety.

Along the Alakahi Stream Trail, there are more than 33 varieties of plants. (See plant list, page 89.)

TOP RIGHT: WATER HYACINTHS AT LILY LAKE *(EICHORINIA CRASSIPES)*. ABOVE: TRAVELERS TREES *(RAVENALA MADAGASCARIENSIS)* BORDER LILY LAKE.

Lily Lake

If, as Henry David Thoreau said, "the lake is the eye of the land," then Lily Lake is the guardian of Onomea Valley. Located at the center of the Garden, the lake now looks as though nature placed it there. In fact, it was dug by hand by Dan Lutkenhouse and three helpers, and the bottom was lined with hand-poured concrete three inches thick to contain the water.

Over 110 species of tropical plants can be seen from the Lily Lake Vista. Among them are giant Queen Victoria water lilies from the headwaters of the Victoria River in Africa; an exotic wi apple tree, believed to be the largest wi apple tree in Hawaii; Madagascar travelers trees, a spectacular fan-shaped variety; and betel nut palms, the source of betel nuts, a narcotic. Purple lotus and papyrus reeds fringe the border of the lake. Nowhere else in the world, be it jungle or botanical garden, is this vast variety of plants displayed.

Lily Lake is home to a large school of beautiful koi fish, some as long as 30 inches. Koi fish are revered in Japan, where they are carefully crossbred to produce prize specimens, sometimes valued in excess of $1000 apiece. Feeding time at Lily Lake is a colorful frenzy, as the fat and happy koi gobble pelletized fish food.

Skirting the lake, the trail passes multi-colored crotons, or Joseph's coat, and pretty impatiens flowers. Ti plants with narrow, red-striped leaves fill the understory, while monsteras with their huge leaves climb into the canopy of tall trees, including giant monkeypod trees. Along this trail are more than 40 different species of plants. (See plant list, page 89.)

ABOVE AND FOLLOWING PAGES:
FROM LILY LAKE VISTA, IT IS SAID
THAT MORE SPECIES OF TROPICAL
PLANTS CAN BE SEEN THAN
ANYWHERE ELSE IN THE WORLD.

BOULDER CREEK
AND COOK PINE TRAILS

Cook Pine Trail

The newest trail in the Garden is the Cook Pine Trail. Whereas most of the Garden trails are shaded by tall trees, the Cook Pine Trail enjoys an abundance of sunshine. Over 48 species of trees and colorful flowering plants that thrive in full sun have been planted here. (See plant list, page 91.)

This trail features a towering Cook pine, named after the famous

RIGHT: A RAINBOW OF COLOR: PURPLE, RED, AND ORANGE HIBISCUS *(HIBISCUS C.V. BLUE BAYOU AND HIBISCUS HYBRIDS)*.
BELOW: THE NEWLY PLANTED HIBISCUS GARDEN.
OPPOSITE: TWO LARGE BOTTLE PALMS *(HYOPHORBE LAGENICAULIS)* AT THE ENTRANCE TO COOK PINE TRAIL.

English navigator Captain James Cook, the first westerner to arrive in Hawaii. Set amidst African tulip trees with their orange blossoms, the Cook pine emerges from the canopy like a rocket to the moon. It is estimated to be 160 feet tall, making it the largest specimen in the islands.

Many tree ferns can be seen here, including the rough tree fern, the black tree fern, and the West Indian tree fern. Numerous blooms of the hibiscus family present a rainbow of colors, ranging from red and yellow to soft purple. Colorful ti plants and unusual stilt root palms line the trail. A noteworthy rare species is the lovely lemon bay rum tree from Trinidad, remarkable for its intoxicating scent.

Cook Pine Trail hosts a variety of Hawaiian endemic plants. The rare palm *Pritchardia schattauerii* is represented; only a few dozen individuals of this species are found today on the Big Island. The loulu palm is also here, a fan palm native to the island of

BELOW: A PORTION OF COOK PINE TRAIL.

LEFT: COLORFUL PLANTS IN THE JUNGLE ALONG COOK PINE TRAIL. BELOW: THE MANY-THORNED TRUNK OF THE FLOSS SILK TREE (*CHORISIA SPECIOSA*).

Molokai. The floss silk tree, with its prominent thorns, will one day be 50 feet tall and bear beautiful blossoms and seeds that look like small footballs. Inside the seeds is a fluffy silk known as kapoc, once used to make life preservers because it floats so well.

Future plans call for the Cook Pine Trail to be extended much farther into the dense jungle on the Garden's upper slopes, allowing visitors to experience a true Amazon-like rainforest setting.

Boulder Creek Trail

On the way to Cook Pine Trail, Boulder Creek Trail crosses Alakahi Stream, which cascades over mammoth boulders. Ordinarily the streambed is dappled with sunshine and mossy rocks, but after heavy rains the stream is transformed into a whitewater torrent. Ferns and flowers grow from hidden nooks in the rocks. Low rock walls which created terraces are evidence of the early settlers who planted taro here. There are more than 16 species of plants along this short, shady trail. (See plant list, page 92.) Nearby, the Garden's historic Portuguese oven is prominently identified.

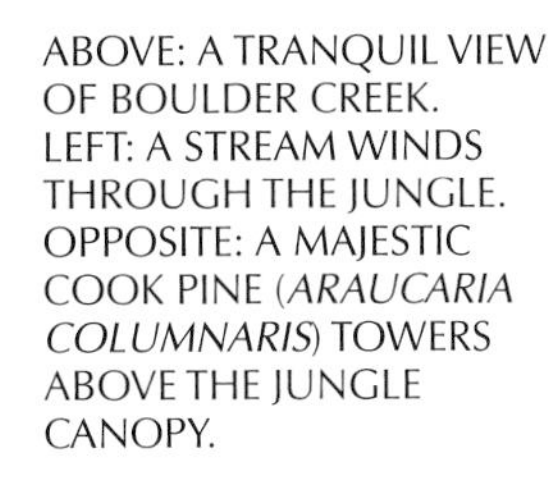

ABOVE: A TRANQUIL VIEW OF BOULDER CREEK.
LEFT: A STREAM WINDS THROUGH THE JUNGLE.
OPPOSITE: A MAJESTIC COOK PINE (*ARAUCARIA COLUMNARIS*) TOWERS ABOVE THE JUNGLE CANOPY.

Lily Lake Trail to Bird Aviaries

More often than not, when you walk past the west side of Lily Lake, the air is filled with the hearty vocalizing of macaws. With their brilliant plumage, these South American birds enhance the Garden's tropical atmosphere. They include a pair of scarlet macaws, a pair of blue and gold macaws, and a pair of hybrid red and green rainbow-colored macaws.

The birds are named after three prominent landscape features: Ono and Mea, for Onomea Bay; Hama and Kua, for the Hamakua Coast; and Hono and Lii, for the Honolii River located in another beautiful valley near Hilo. Macaws can live for over a hundred years, and a breeding program is planned for these young birds to help insure the survival of their species.

Along the trail are several unique red sealing wax palms, so named by the British in Singapore because their trunks were the color of the brilliant red wax then used to seal

INTELLIGENT AND PLAYFUL, MACAWS CAN BE AFFECTIONATE, AND THEY CAN ALSO LEARN TO SPEAK WELL. THEY LIVE AS LONG AS 100-150 YEARS.

BORN AND RAISED IN HAWAII, THE GARDEN'S MACAWS INCLUDE THREE PAIR: *ARA ARARAUNA* (BLUE AND GOLD) FROM SOUTH AMERICAN JUNGLES; *ARA MACAO* (SCARLET) FROM THE CARIBBEAN AND MEXICO; AND *SHAMROCK*, A HYBRID CROSS BETWEEN A SCARLET MACAW AND A MILITARY MACAW (*ARA MILITARIS*).

important documents. A very rare tall specimen of this palm was donated to the Garden by Margaret Hirose, an ardent plant lover and long-time Hilo resident whose lifelong dream was to establish a botanical garden. It is the only red sealing wax palm in the entire state of Hawaii known to produce seeds. A plaque in memory of the late Margaret Hirose, who also donated many other specimens to the Garden, is placed at the foot of this beautiful tree.

Along this trail you will see more than 40 species of plants. (See plant list, page 92.)

Bromeliad Hill

Farther along the Outer Lake Trail is a small hill covered with a vast variety of bromeliads. These plants are true tropicals, relatives of the pineapple that made Hawaii famous. There are more than 1400 recognized species of bromeliads in the world, most of which originated in the South American jungle. More than 80 varieties and species are represented in the Garden. (See plant list, page 89.)

Bromeliads are epiphytes, meaning air plants, and they rely on rainwater to provide their nutrients.

OPPOSITE: SMALL PURPLE BLOOMS SURROUND THE ROOTS OF A MONEY TREE (*DRACAENA MARGINATA*). RIGHT: JUNGLE TRAIL PASSING BY LILY LAKE. BELOW: VARIEGATED RED PINEAPPLE (*ANANAS BRACEATUS, VAR. STRIATUS*).

ABOVE AND LEFT: BROMELIADS (*GUZMANIA*). THE GARDEN FEATURES AN EXTENSIVE COLLECTION OF BROMELIADS (INCLUDING PINEAPPLES).

THE GARDEN HOUSES ONE OF THE LARGEST COLLECTIONS OF BROMELIADS IN THE STATE OF HAWAII.

Pools of water collect in the crown of the plants, along with dirt and leaves which provide additional nourishment.

Some of the Garden's finest bromeliads are found on the roots and trunk of a giant fallen banyan tree. This former matriarch of the valley was a century old, but a storm toppled the behemoth several years ago. It is a fitting requiem that in death, the tree supports a living bromeliad collection.

Orchid Garden

Close to Bromeliad Hill lies the Orchid Garden, which displays many distinct species. (See plant list, page 89.) Fabulous flowers sporting every

AN EVER-CHANGING COLLECTION OF COLORFUL ORCHIDS IS ON DISPLAY NEAR THE SHORES OF LILY LAKE.

ABOVE: ORANGE ONCIDIUM ORCHID (*ORCHID HYBRID*). RIGHT: MOTH ORCHID (*PHALAENOPSIS VAR.*). OPPOSITE: LADY'S SLIPPER ORCHID (*PAPHIOPEDILUM VAR.*).

ABOVE: THE WINGS OF LESSER FLAMINGOS (*PHOENICONAIAS MINOR*) ARE MOTTLED WITH DEEP CRIMSON, AND FEATURE BLACK UNDERFEATHERS.

THE NATIVE HOME OF LESSER FLAMINGOS IS LAKE MANYARA, WHICH LIES IN THE RIFT VALLEY OF TANZANIA, AFRICA.

shade of pastel, including tiger orchids, butterfly orchids, and cattleyas, all populate this corner of the Garden. The orchids on view are only a few of the estimated 25,000 known species worldwide. In addition, more than 60,000 new varieties have been created through hybridization.

Flamingo and Duck Pond

Nearby, nestled in the jungle, a small pond is home to a family of pink Lesser flamingos and Mandarin ducks. The Lesser flamingos are natives of Lake Manyara in Tanzania, and they sport a more delicate shade of pink than their more familiar relatives often seen in zoos.

Renowned in China and Japan for thousands of years, Mandarin ducks are gregarious birds with strong bonds to their mates. In ancient China, pairs of Mandarin ducks were given as wedding gifts.

Near the Flamingo Pond are an abundance of food plants, such as thousand-fingered bananas and rare red bananas. The cocoa tree occca-sionally produces a pod of chocolate beans. Another example is the Moreton Bay chestnut from Australia, which has recently been found by King's College of London to be of possible use in a cure for AIDS. All

ABOVE: MALE MANDARIN DUCKS FEATURE GORGEOUS PLUMAGE, WITH ORANGE-GOLD "SAILS" ALONG THEIR FLANKS. RIGHT: THE THOUSAND-FINGERED BANANA (*MUSA CHILIOCARPA*) BOASTS UP TO 1,000 SMALL BANANAS ON A SINGLE STALK.

these valuable plants come from the tropics, underscoring the importance of the rainforests to mankind.

Along the trail between the Flamingo Pond and Fern Circle are more than 50 species of plants. (See plant list, page 90.)

Torch Ginger Forest

From here visitors enter a large grove of torch ginger. Introduced to Hawaii from the island of Mauritius in the Indian Ocean, torch gingers thrust their huge flower heads skyward from long, thick red stalks. The showy, waxen red cones bloom from May to July and can grow up to eight inches in

ABOVE: HONEYCOMB GINGER (*ZINGIBER SPECTABILE*).
LEFT: A SPECTACULAR RED TORCH GINGER BLOSSOM (*NICOLAIA ELATIOR*).

THREE OF THE MANY GINGERS IN THE GARDEN'S COLLECTION.
ABOVE LEFT: BLUE GINGER (*DICHORISANDRA THYRSIFLORA*).
ABOVE RIGHT: RED GINGER (*ALPINIA PURPURATA*).
BELOW RIGHT: PINK GINGER (*ALPINIA PURPURATA*).
OPPOSITE: RARE ROUGE PUFF BLOSSOM (*BROWNEA MACROPHYLLA*).

diameter. The plants, which reach heights of 20 feet, have enormous leaves that shade the path below.

Frequently orange-red brownea blossoms, also called rouge puffs, burst out of their buds on the naked trunk of the brownea tree. Lucky is the visitor to spy this extraordinary blossom, which resembles a huge orange powder puff.

Wild Birds in the Garden

The Garden has an interesting variety of wild birds who make their home in the jungle. The cardinal, common yet quite beautiful, is often seen as a flash of red amidst the Garden's deep green foliage. The Hawaiian hawk soars above the Garden, seeking its next meal or just enjoying its flight. The beautiful red tail tropic bird nests on the cliffs and is frequently seen flying high over Onomea Bay.

The *'Apapane* and the *'Amakihi* are examples of native Hawaiian birds that lived at one time in the lowland tropical rainforests. Over the last thousand years mankind has devastated these forests along the Hamakua Coast. Many native bird species became extinct when their forest homes were destroyed, but others, like the *'Apapane* and *'Amakihi,* still survive precariously in the remaining rainforests high on the slopes of Mauna Kea and Mauna Loa. By expanding and preserving the rainforest surrounding Onomea Bay, the Garden hopes to entice these and other native birds to again find a home here.

TOP: WHITE TROPIC BIRD.
MIDDLE: RED APAPANE, RED OHIA.
BOTTOM, LEFT TO RIGHT: HAWAIIAN HAWK, NESTING WHITE TROPIC BIRD, AND AMA KIHI BIRD WITH A LOBELIA FLOWER.

Giant Fern Circle

Approaching the Giant Fern Circle, we are near the heart of the Garden. Giant cycads, believed to be the first plants that colonized our planet, thrive in this moist rainforest climate. Giant tree fern fronds reach for the sky, sword ferns wreathe the path, and primitive cycads complete the prehistoric atmosphere. Watch for the Flame of the Forest high up in the banyan tree — a spectacular red flowering vine whose seeds were brought to the Garden from Australia by the Lutkenhouses.

Palm Vista Trail

Visitors come from all over the world to view the Garden's extraordinary collection of palms. Members of the Royal Palm Society of London excitedly inspected the rare Peruvian palm *Euterpe precatoria* here. The Garden is home to nearly 200 species of palms: fan, fish-tail, sago, date, and betel nut palms are just a few that can be seen from the Palm Vista. Other interesting palms include the wanga palm from Malaysia, the fastest-growing palm in the world; and the orange areca palm, with its fascinating orange seeds.

ABOVE LEFT: EXOTIC FERN FRONDS IN THE GIANT FERN CIRCLE.
ABOVE RIGHT: TALL AUSTRALIAN TREE FERNS (*ALSOPHILA AUSTRALIS*).
INSET: LOVELY PINK CALADIUM LEAVES (*CALADIUM HORTULANUM*).

Along the Palm Vista Trail, from the Fern Circle to the bottom of the steps near the Palm Jungle, you will see more than 50 species of plants and trees—including the giant jackfruit tree with its enormous colorful fruit, sometimes more than two feet long! (See plant list, page 90.)

Palm Jungle Trail

If the Giant Fern Circle is the heart of the Garden, then the Palm Jungle is the soul. Here a forest of towering Alexandra palms creates the feeling of a cathedral, and visitors often fall silent as they enter. Originally from Australia, Alexandra palms grow profusely in valleys along the Hamakua Coast. They have long, creamy flower clusters that appear below the fronds. These flowers produce huge clusters of seeds which turn from green to a beautiful red. Then they fall and carpet the jungle

ABOVE: ALEXANDRA PALM WITH RED SEEDS (*ARCHONTOPHOENIX ALEXANDRAE*). RIGHT: COCONUT PALM (*COCOS NUCIFERA*) AND AN UNUSUAL RED-COLORED ROYAL POINCIANA SABAL TREE (*DELONIX REGIA SABAL SP.*). OPPOSITE PAGE: CARIBBEAN ROYAL PALM (*ROYSTONEA OLERACEA*).

ABOVE: THE SERENITY OF UPPER BOULDER CREEK.
INSET: WHITE TRUMPET VINE (*THUNBERGIA GRANDIFLORA*).

floor to produce hundreds of tiny new palm tree seedlings.

After wending your way through the Palm Jungle, you will find Onomea Stream descending from the mountains through this cool glade of palms on its way to the sea. Onomea Stream has created the crown jewel of the Garden, Onomea Falls.

Onomea Falls

This spectacular three-tiered waterfall is often claimed to be the most beautiful in Hawaii. It was discovered far back in the jungle by Dan Lutkenhouse, years after work on the lower Garden began. One day he decided to hack his way through the jungle alongside the stream, and much to his delight he found the magnificent waterfall.

Onomea Falls is set amidst the natural forest of palms and ferns. Exotic mosses grow on the surrounding rocks and trees. Small fish and prawns thrive in the clear, cool water.

Gazing at Onomea Falls from the viewing bridge, visitors are treated to an experience of unmatched natural beauty. There is a feeling of deep peace and serenity here, as well as a sense of the power and abundance of nature.

ABOVE: ONOMEA FALLS IS RENOWNED THE WORLD OVER FOR ITS STUNNING BEAUTY.
LEFT: ROOT STRUCTURE OF AN ALEXANDRA PALM (*ARCHONTOPHOENIX ALEXANDRAE*).

ABOVE: HANGING HELICONIA (*HELICONIA COLLINSIANA*). HELICONIAS ORIGINATE IN THE JUNGLES OF CENTRAL AND SOUTH AMERICA, AS WELL AS SOME SOUTH PACIFIC ISLANDS.

Heliconia Trail

After leaving the Palm Jungle, visitors wander along the spectacular Heliconia Trail where more than 80 different plant species grow. (See plant list, page 91.) The Garden has one of the finest collections of heliconias in the United States. In this protected environment, heliconias grow more beautifully than in their native rainforest jungle. Colorful, striking flowerheads in a multitude of bizarre shapes and sizes arise from clumps of large, oval leaves. The plants range from 2 to 20 feet high.

Once classified with bananas, heliconias are now considered a separate family, *Heliconiaceae*. The dramatic, colorful parts of the plant are not actually flowers, but rather highly modified leaves called bracts. The heliconia's inconspicuous true flowers are located inside the bracts.

The best time to see heliconias in bloom is from May through August;

LEFT: GOLDEN TORCH HELICONIA (*HELICONIA HYBRID*).

ABOVE: HELICONIA (*HELICONIA BOURGAEANA*). LEFT: RED LOBSTER CLAW HELICONIA (*HELICONIA CARIBAEA*). THERE ARE MORE THAN 450 DIFFERENT SPECIES, VARIETIES AND HYBRIDS OF HELICONIAS.

THE GARDEN IS HOME TO 82 SPECIES AND VARIETIES OF HELICONIAS, AND THE COLLECTION CONTINUES TO GROW.
ABOVE LEFT: HANGING LOBSTER CLAW HELICONIA (*HELICONIA ROSTRATA*).
ABOVE RIGHT: RED HANGING HELICONIA (*HELICONIA PENDULA*).
BELOW RIGHT: RAINBOW HELICONIA (*HELICONIA WAGNERIANA*).

however, various species can be viewed throughout the year. Among the most striking varieties is the hanging lobster claw—a cluster of glorious, vibrant red flowerheads tipped with green and yellow. Another species, *Heliconia metallica xosaensis*, was brought to the Garden from Costa Rica shortly before the section of rainforest it came from was destroyed.

New species of heliconias are being discovered each year as botanists explore deeper in the jungle. The Garden constantly adds to its collection, and is recognized as one of the foremost centers for the cultivation and preservation of heliconias in the world.

FAR LEFT: *HELICONIA RAMONENSIS.*
LEFT: HELICONIA (*HELICONIA CHAMPNEIANA C.V. SPLASH*).
BELOW: RED CHRISTMAS HELICONIA (*HELICONIA ANGUSTA*).

ABOVE: GARDENIA REMYI BLOSSOM, NATIVE TO HAWAII.
RIGHT: BANYAN TREE (*FICUS SP.*).
BELOW: GREEN STRIPE BAMBOO (*BAMBUSA VULGARIS, VAR. VITTATA*).

Banyan Canyon

Where Onomea Stream glides over another small waterfall, the cool, dark shade of Banyan Canyon beckons. Here a tall banyan tree clings to the wet rocks with a million roots. A species of fig, the banyan sends tiny roots down from its larger branches; these roots soon grow and form new tree trunks, large enough to provide additional support and nourishment. The roots dangle in the air and sweep the ground before taking hold in the soil.

Nearby are unique palms with blue-green trunks, a bamboo forest, and a rare native Hawaiian gardenia known as *remyi*. A huge stump bears witness to the grandeur of a giant fallen banyan that now supports part of the Garden's bromeliad collection.

Vista Points

Following the yellow arrows, visitors find many small side trails. Wiapple Vista, Coconut Vista, and Monkeypod Vista all offer the chance to sit a spell on a shady bench contemplating Onomea Stream and the ocean waves crashing below. This is the perfect place to breathe the pure, scented air and absorb the Garden's relaxing atmosphere one last time before heading out of this enchanted valley.

VISITORS ENJOY A LAST LOOK AT THE GARDEN'S SERENE OCEAN VISTAS.

PART 3

Growing Towards the Future

There is always music amongst the trees
in the garden, but our hearts must be
very quiet to hear it.
—Minnie Aumonier

With over 2,000 species of tropical plants, the Hawaii Tropical Botanical Garden provides a home for unique and endangered flora collected from the jungles of the tropical world. Many of the exotic tropical plants and trees that thrive in the Garden are rapidly becoming extinct in the wild.

Scientists estimate that the world has 20 to 50 years to collect and study the flora on our planet. After that, the burning, clear-cutting, and overuse of forests will have destroyed 90 percent of the species now alive.

The Garden acts as a living seed bank to insure the future of these species. Seeds from rare tropical plants and trees are distributed by the Hawaii Tropical Botanical Garden to other botanical gardens and collectors from around the world. As rainforest destruction continues, the Garden is increasing its collection activities to ensure that future generations will be able to see and appreciate these natural wonders.

ABOVE: YELLOW SHOWER TREE (*CASSIA SP.*). PRECEDING PAGE: THE TRAIL PAST AGAVE HILL. INSET: RED FLOWERING BANANA (*MUSA COCCINEA*).

The Garden's by-laws have been carefully designed to preserve the natural environment in perpetuity. The land is held in trust and can never be sold, and the Garden's board of directors is charged with seeing that the work of the Garden continues. Nowhere in the United States is there a location offering a similar combination of benefits to science and environmental studies, plus a natural setting for the enjoyment and education of the general public.

Educational Programs

The Garden's main purpose is to serve as an educational center for all visitors, and most especially for teaching young children. The trails themselves provide an "outdoor classroom" of living plants. Signs along the trails identify plants and trees by common and botanical names, as well as country of origin. Some relate legends connected with particular plants and tell the stories of Garden landmarks.

For many years the Garden's Outdoor Classroom Program has invited schoolchildren and community groups to visit the Garden in order to learn more about the environment and the importance of preserving tropical rainforests. Nearly 300 students per month participate in the Outdoor Classroom Program. The program is free of charge to children, and the Garden absorbs all expenses related to this important educational program.

An open-air Educational Pavilion provides chairs, tables, and benches for the use of visiting school and community groups. Lectures are given here by Garden staff and educators. The Educational Pavilion is also used as an activity and exhibit site.

TOP: DAN LUTKENHOUSE UNDER A WI APPLE TREE (*SPONDIAS CYTHEREA*). ABOVE: LARGE PELICAN FLOWER (*ARISTOLOCHIA GIGANTEA*).

TOP: TEMPLE FLOWERS (*CLERODENDRUM PANICULATUM*).
ABOVE LEFT: BEGONIA LEAF.
ABOVE RIGHT: CROTON (*CODIAEUM VARIEGATUM*).
LEFT: YELLOW SHOWER TREE BLOSSOM (*CASSIA FISTULA*).

The Garden is Growing!

In 1989 Dan and Pauline Lutkenhouse purchased an additional 40 acres of land, including all of the beautiful tropical rainforest surrounding Onomea Bay. In the near future the Garden may be expanded to include some of this new land. Plans include creating a new forest of tropical trees, palms, and flowers, many of which will be Hawaiian native plants.

A new Agricultural Center is planned as well. Serving as the Garden headquarters, the Agricultural Center will house offices, a small museum of Hawaiian artifacts and rare antiques donated by Dan and Pauline, and a beautiful gift shop selling plants, flowers, and a vast variety of books about nature and plants.

With the aid of engineers, Dan Lutkenhouse is designing a small surface tram which will eventually transport visitors to the Garden. The proposed tram ride will be an exciting feature, with spectacular panoramic views of the jungle during the descent into the valley.

ABOVE: PAPYRUS AND A VAST VARIETY OF PLANT SPECIES CAN BE SEEN GROWING ALONGSIDE THIS TRAIL NEAR LILY LAKE. RIGHT: THOUSANDS OF SCHOOLCHILDREN VISIT THE GARDEN EACH YEAR THROUGH THE OUTDOOR CLASSROOM PROGRAM.

Sources of Financial Support

For the first 10 years of the Garden's existence, Dan and Pauline Lutkenhouse personally funded the acquisition of the land and the Garden's operating expenses, at an approximate cost of $2 million.

Currently the Garden is generating sufficient monies to pay operating and maintenance expenses, which are approximately $600,000 annually. A small cash surplus is being created, to be used for future development of the Garden and for construction of the new Agricultural Headquarters.

From time to time, foundations and businesses have donated funds to the Garden, including the C. Brewer Charitable Foundation, The Hearst Foundation, Inc., G. N. Wilcox Trust, Frear Eleemosynary Trust, James and Abigail Campbell Foundation, The First Hawaiian Foundation, and Orchid Isle Properties. In addition, individual members have voluntarily donated funds ranging from $5 to $10,000, expressing their appreciation of what the Garden represents.

An Endowment Program and a Children's Educational Trust are being established. The Lutkenhouse Trust is expected to be the first of such endowments. As the Garden expands and its value to the world's people becomes more widely known, it is anticipated that more people will enjoy the opportunity of supporting the Garden with their endowments.

Hawaii Tropical Botanical Garden is a non-profit public charity IRS Section 501(c)(3). All donations, grants, and endowments are tax deductible to the extent allowable by law.

ABOVE: THE FOUNDERS, DAN AND PAULINE LUTKENHOUSE, AND THE HISTORIC CHURCH WHICH SERVED AS THE GARDEN HEADQUARTERS, MUSEUM AND VISITOR CENTER UNTIL IT WAS DESTROYED BY FIRE IN 1988.
LEFT: THE BURNT CHURCH AND TEMPORARY VISITOR CENTER AND GIFT SHOP.

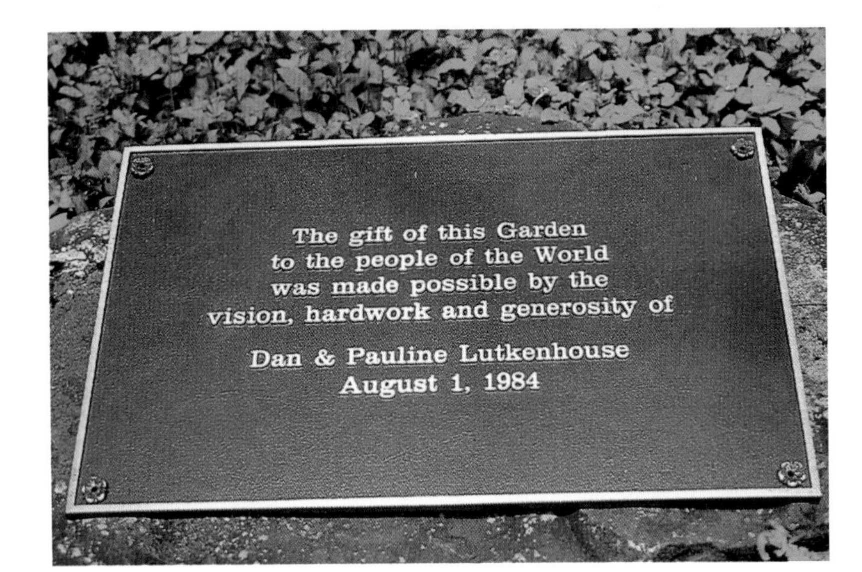

TOP: A BRONZE PLAQUE COMMEMORATES THE LUTKENHOUSES' GIFT OF THE GARDEN TO THE WORLD.
ABOVE: BEAUTIFUL RED HIBISCUS DELIGHT GARDEN VISITORS.

Garden Memberships

You too can help protect the beauty and wonders of nature by becoming a member of the Garden! Garden members make an important contribution to preserving this amazing natural and educational resource for future generations.

Annual membership entitles you to receive the Garden Newsletter and to visit the Garden without charge. Benefactors, Life Members, and Patrons will receive recognition on a plaque within the Garden and may also select a tree to be dedicated in the name of their choice.

The Hawaii Tropical Botanical Garden is a precious jewel that belongs to all of us. Supporting the Garden connects you to worldwide efforts to study and preserve topical flora. The Garden attracts and inspires visitors from around the world. Together, we can help the Garden grow into a lasting treasure.

Our Garden Family of Visitors

From three guests on opening day in 1984, the Garden's family of visitors has grown steadily—now averaging 200 people a day, for a total of approximately 50,000 visitors per year. Children under 16 years of age are admitted free, and more than 5,000 children visit the Garden each year.

People throughout the world are becoming aware of the dire need to preserve rainforests and tropical plants and trees. This interest in conservation is bringing more and more people to the Garden. Yet even on the busiest days, the Garden is never overcrowded. Visitors will always find a peaceful experience of nature here. If the number of visitors continues to increase, reservations may eventually be required to preserve the serenity of the Garden experience.

The Hawaii Tropical Botanical Garden's reputation has spread, and visitors are coming to the Garden from virtually every country in the world. Movies and videos are frequently filmed in the Garden by producers from Hollywood and from many countries of the world. *National Geographic, Smithsonian,* travel and garden magazines, and many newspapers have written editorials about this spectacular garden.

Many visitors exclaim, "This is what I dreamed Hawaii would be like!" A visit to Hawaii Tropical Botanical Garden is a one-of-a-kind tropical jungle experience. For many, this lush valley is the highlight of their trip to Hawaii.

Visitor Comments

I've seen many tropical rainforests, and I have never seen so much beauty in one place! The best tropical garden I have ever seen!
Gudrun Kruss, Elmshorn, Germany

The most glorious botanical display I have ever seen! Thank you for your dedication in preserving this wonderful valley.
Dee Vandercook, Grover City, CA

This is heaven on earth . . . a true paradise found. Spectacular!!
Loretta and David Koleos,
Marietta, GA

We feel it was a once-in-a-lifetime experience. Truly a labor of LOVE!
Richard J. Hilkin, Woodinville, WA

The scenery was breathtaking! Mere words cannot express nature at its best. This is what I had always pictured when thinking of Hawaii. Truly paradise!
Rose Mary Hamilton, Mishawaka, IN

Absolutely beautiful! Very tranquil, immensely enjoyable. Thank you for providing everyone with a glimpse of "Old Hawaii."
Kathleen Dirickson, San Francisco, CA

Hawaii Tropical Botanical Garden Plant List

On the following pages you will find a partial list of the more than 2,000 species of tropical plants on display at the Hawaii Tropical Botanical Garden. The listings are arranged by trail and section of the Garden.

Plants Along Trail 1 to the Ocean

COMMON NAME	BOTANICAL NAME	ORIGIN
Licuala	Licuala Grandis	New Hebrides Is.
Licuala	Licuala Spinosa	Thailand
Painted Droptongue	Aglaonema Crispum	Asia/Philippines
Amazon Lily	Eucharis Grandiflora	Colombia/Peru
Hapu'u	Cibotium Glaucum	Hawaii
Snowflower	Spathiphyllum Floribundum	Colombia
Black Anthurium	Anthurium Watermaliense	Colombia
Anthurium	Anthurium Andraeanum	Colombia
Bat Flower	Tacca Chantrieri	Southeast Asia
Calathea	Calathea Picturata	Brazil
Lady Palm	Rhapis Excelsa	China/Japan
Calathea	Calathea Ornata	Tropical America
Spiral Flag	Costus Malortieanus	Central America
Mules Foot Fern	Angiopteris Evecta	Japan/Australia/Madagascar
Kris Plant	Alocasia Sanderana	Philippines
African Mask Plant	Alocasia Micholitziana	Philippines
Cardboard Plant	Zamia Pumila	Florida/W. Indies/Mexico
Cyrtosperma	Cyrtosperma Johnstonii	Solomon Islands
Impatiens	Impatiens Wallerana	Tanzania/Mozambique
Chinese Fan Palm	Livistona Chinensis	China/Japan
Queen Sago	Cycas Circinalis	Old World Tropics
Metal Palm	Chamaedorea Metallica	Mexico
Palm	Rhapis Minor	Thailand
Cast Iron Plant	Aspidistra Elatior	China
Bamboo Palm	Chamaedorea Sp.	Mexico/S. America
Marica/Walking Iris	Neomarica Gracilis	Tropical America
Australian Tree Fern	Sphaeropteris cooperi	Australia
Alexandra Palm	Archontophoenix Alexandrae	Australia
Fishtail Palm	Caryota Cumingii	Philippines
Tricolor Money Tree	Dracaena Marginata c.v. Tricolor	Madagascar
Dragon Bones	Euphorbia Lactea	India
Japanese Sedge	Carex Morrowii Variegata	Japan
Kings Mantle	Thunbergia Erecta	Tropical Africa
Crown of Thorns	Euphorbia Milii	Madagascar
Weddel Palm	Microcoelum Weddelliana	Brazil
Rose Grape	Medinella Magnifica	Philippines
Pink Ginger	Alpinia Purpurata c.v Eileen McDonald	Hort/Pacific Is.
Bird Nest Anthurium	Anthurium Hookeri	Guiana
Tree Maidenhair Fern	Didymochileana Truncatula	India, Africa, Malaysia
Latanier Latte	Verschaffeltia Splendida	Seychelles Islands
Guzmania	Guzmania Lingulata	Tropical America
Painted Drop Tongue	Aglaonema Crispum	Philippines
Areca Palm	Areca Vestiaria	Moluccas/Indonesia
Manila Palm	Veitchia Merrillii	Philippines
Parlor Palm	Chanaedorea Elegans	Mexico/Guatamala
Pinanga Palm	Pinanga Kuhlii	Java/Sumatra
Birdsnest Fern	Asplenium Nidus	Polynesia
Spathiphyllum	Spathiphyllum Sp.	South America
Spider Lily	Crinum Asiaticum	Tropical Asia
Coconut	Cocos Nucifera	World Tropics
Kou	Cordia Subcordata	Africa to Polynesia

Cardinal (*Cardinale*).

Hibiscus (*Arnottianus*).

COMMON NAME	BOTANICAL NAME	ORIGIN
Lauae Fern	Polypodium Scolopendria	Old World Tropics
Monkey Pod	Samanea Saman	Tropical America
Common Apé	Xanthosoma Robusta Var. Robustum	Mexico/Central America
Hala	Pandanas Sp.	South Pacific
Ironwood	Casurina Equisetifolia	Australia
Beach Naupaka	Scaevola Frutescens	South Pacific
Mango	Mangifera Indica	India/Asia
Guava	Psidium Guajava	Tropical America
Indonesian Ginger	Tapeinochilus Ananassae	Malaysia

Plants Along the Oceanfront Trail

COMMON NAME	BOTANICAL NAME	ORIGIN
Hala/Screwpine	Pandanus Sanderi	S. Pacific/Timor
Breadfruit	Artocarpus Altilis	Polynesia
Champaca Tree	Michelia Champaca	Java
Blue Trumpet Vine	Thunbergia Grandiflora	India
Jaboticaba	Myrciaria Cauliflora	Brazil
Magnolia	Magnolia Grandiflora	United States
Mammay Apple	Mammea Americana	West Indies
Croton	Codiaeum Variegatum	Malay Peninsula/Pacific Is.
Cacao	Theobroma Cacao	Central/S. America
Sea Putat	Barringtonia Asiatica	Madagascar
Whaleback	Molineria Recurvata	Asia/Australia
Baby Doll Ti	Cordyline Terminalis	Horticultural
Dwarf Hau	Hibiscus Tiliaceus	World Tropics
Fountain Bush	Russellia Equisetiformis	Mexico
Be Still Tree	Thevetia Peruviana	Tropical America
Spider Lily	Crinum Asiaticum	Tropical Asia
Mauritus Hemp	Furcraea Foetida	South America
Sabal Palm	Sabel Sp.	Mexico
Asparagus Fern	Asparagus Densiflorus Var. Sprengeri	South Africa
Royal Poinciana	Delonix Regia	Madgascar
Sea Grape	Coccaloba Uvifera	Florida/S. America
Beach Morning Glory	Ipomoea Pes Caprae	World Tropics
Kamani	Calophyllum Inophyllum	Asia/Pacific
Hala/Screwpine	Pandanas Odoratissimus	South Pacific
False Kamani	Terminalia Catappa	Malay Peninsula
Soft Tip Agave	Agave Attenuata	Mexico
Little Club Moss	Selaginella Sp.	World Tropics
Alexandra Palm	Archontophoenix Alexandrae	Australia

Plants Along Alakahi Stream Trail

COMMON NAME	BOTANICAL NAME	ORIGIN
Calathea	Calathea Leopardina	Tropical America
Zebra Plant	Calathea Zebrina	Brazil
Calathea	Calathea Ornata	Tropical America
Calathea	Calathea Albertii	Unknown
Calathea	Calathea Bella	Brazil
Prayer Plant	Maranta Leuconeura	Brazil
Heliconia	Heliconia Stricta c.v. Dwarf	Jamaica/Ecuador
Calathea	Calathea Rufibarba	Brazil
Warabi/Paco	Diplazium Esculentum	Polynesia
Yellow Ginger	Hedychium Flavescens	India
Kukui Nut	Aleurites Moluccana	Southeast Asia
Pothos	Epipremnum Aureum	Southeast Asia
Calathea	Calathea Warscewiczii	Costa Rica
Purple Waffle Plant	Hemigraphis Colorata	Malaysia
Imperial Philodendron	Philodendron Speciosum	Brazil
Amherstia	Amherstia Nobilis	Burma
Heliconia	Heliconia Mettalica	Colombia
Heliconia	Heliconia Imbricata	Costa Rica
Heliconia	Heliconia Champneiana	Central America
Dwarf Flowering Banana	Musa Velutina	Assam
Parrott Flower	Heliconia Psittacorum	Brazil
Anthurium	Anthurium Andraeanum	Colombia
Mango	Mangifera Indica	India
Pink Ginger	Alpinia Purpurata	Horticultural
Dwarf Date Palm	Phoenix Roebelinii	Vietnam/Laos
Peter Buck Ti	Cordyline Terminalis	Horticultural
Baby Doll Ti	Cordyline Terminalis	Horticultural
Princess Palm	Dictyosperma Album	Mascarene Is.
Rhopaloblaste	Rhopaloblaste Ceramica	Indonesia
Flowering Banana	Musa Coccinea	Vietnam
Bottle Palm	Hyopyorbe Lagenicaulis	Mascarene Is.
Triangle Palm	Neodypsis Decaryi	Madagascar
Wandering Jew	Zebrina Pendula	Mexico

Plants Along Lily Lake Trail to Bird Aviaries

COMMON NAME	BOTANICAL NAME	ORIGIN
Blood Leaf	Iresine Herbstii	Brazil
Papyrus	Cyperus Papyrus	North Africa
Licuala	Licuala Peltata	Burma/India
Bromeliad	Aechmea Mexicana	Mexico
Yucca	Yucca Elephantipes	Mexico
Dwarf Mondo Grass	Ophiopogon Japonicus	Japan/Korea
Japanese Sago	Cycas Revoluta	China/Japan
Croton	Codiaeum Variegatum	Malay Pen./Pacific
Mountain Orchid	Spathiglottis Plicata	Malaysia
Orange Costus	Costus Igneus	Northeast Brazil
Varigated Spider Lily	Crinum Sp.	
Fern Leaf Aralia	Polyscias Filicifolia	South Pacific
Ming Aralia	Polyscias Fruticosa	India/Polynesia
Thatch Palm	Thrinax Morrisii	West Indies
Giant Alocasia	Alocasia Odora	Southern Asia
Sealing Wax Palm	Crytostachys Renda	Malaysia
Macarthur Palm	Ptychosperma Macarthurii	Australia
Queen Palm	Syagrus Romanzoffianum	Brazil
Blue Latan Palm	Latania Loddigesii	Mascarene Is.
Silver Thatch Palm	Coccothrinax Argentea	Hispanola
Latanier Latte	Verschaffeltia Splendida	Seychelles Is.
Latanier Feulle	Phoenicophorium Borsigianum	Seychelles Is.
Blue Ginger	Dichorisandra Thyrsiflora	Brazil
Lucuba Palm	Chrysalidocarpus Lucubensis	Madagascar
Cochliostema	Cochliostema Odoratissimum	Costa Rica
Socratea	Socratea Durissima	Central America
Calyptrogyne	Calyptrogyne Sp.	Mexico/Guatamala
Christmas Heliconia	Heliconia Angusta	Brazil
Dumbcane	Dieffenbachia Maculata	Brazil
Heliconia	Heliconia Collinsiana	S. Mexico/Nicaragua
Heliconia	Heliconia Aurantiaca	Mexico/Panama
Lobster Claw	Heliconia Caribaea	West Indies
Heliconia	Heliconia Richardiana	Venezuela/Brazil
Dracaena	Dracaena Thalioides	Tropical Africa

African tulip tree blossom (*Spathodea campanulata*).

Plants Found at Bromeliad Hill

COMMON NAME	BOTANICAL NAME	ORIGIN
Summer Torch	Billbergia Pyramidalis	Brazil
Pink Quill	Tillandsia Cyanea	Ecuador
Pineapple	Ananas Comosus	Brazil
Dwarf Pineapple	Ananas Nanus	Brazil
Variegated Pineapple	Ananas Bracteatus Var. Striatus	Brazil
Heart of Flame	Bromelia Balansae	Argentina
Blushing Bromeliad	Neoregelia Carolinae	Brazil
Bromeliad	Aechmea Mulfordii	Southern Brazil
Pineapple	Ananas Comosus	Brazil

Orchid Garden

COMMON NAME	BOTANICAL NAME	ORIGIN
Arundina	Orchid Species	South America
Cattleya		
Dendrobium		
Epidendrum		
Miltonia		
Odontoglossum		
Oncidium		
Phalaenopsis		
Spathoglottis		
Vanda		
Vanilla		

Plants Near the Flamingo and Duck Pond, and Past the Torch Ginger Forest

COMMON NAME	BOTANICAL NAME	ORIGIN
Prince Kuhio Vine	Ipomoea Horsfalliae	West Indies
Ponytail Plant	Beaucarnea Recurvata	Mexico
Sealing Wax Palm	Cyrtostachys Renda	Malaysia/Pacific
Temple Flower	Clerodendrum Paniculatum	Southeast Asia
Snowflake Plant	Trevesia Palmata	India/China
Iranian Sweet Lime	Citrus Aurantifolia	Iran
Thousand Finger Banana	Musa Chiliocarpa	Malaysia
Honeycomb Ginger	Zingiber Spectabile	
Ice Green Calathea	Calathea Cylindrica	Brazil
White Calathea	Calathea Burle Marxii	Brazil
Ice Blue Calathea	Calathea Burle Marxii	Brazil
Hidden Lily	Curcuma Elata	Burma
Tree Maidenhair Fern	Didymochlaenna Truncatula	Africa/India
Lizard Plant	Alocasia Portei	Philippines
Areca Palm	Chrysalidocarpus Lutescens	Madagascar
Betel Nut Palm	Areca Catechu	Malay Peninsula
Fishtail Palm	Caryota Mitis	Burma/Philippines
Red Crinum	Crinum Amabile	Sumatra
Travelers Tree	Ravenala Madagascariensis	Madagascar
Orchid Ginger	Alpinia Mutica	Malay Peninsula
Hanging Lobster Claw	Heliconia Rostrata	Peru
Cacao	Theobroma Cacao	Central/S. America
Golden Trumpet Vine	Allamanda Cathartica	South America
Caricature Plant	Graptophyllum Pictum	New Guinea
Agave	Agave Geminiflora	Mexico
Candelabra Aloe	Aloe Arborescens	South Africa
Mauritius Hemp	Furcraea Foetida	South America
Aloe Vera	Aloe Barbadensis	Mediterranean Region
Variegated Agave	Agave Americana Marginata	Mexico
Sisal	Agave Sisalana	Mexico
Variegated Agave	Agave Angustifolia Var. Marginata	Mexico
Cycad	Dioon Edule	Mexico
Pikake	Jasminum Sambac	Asia
Cannon Ball Tree	Couroupita Guianensis	Guianas
Chestnut Vine	Tetrastigma Voinieranum	Laos
Cotton Rose	Hibiscus Mutabilis	Southern China
Kapok Tree	Ceiba Pentandra	Ceylon/Java
Kikania Lei	Solanum Aculeatissimum	Trop. America
Variegated Mauritius Hemp	Furcraea Foetida c.v. Mediopicta	South America
Persian Shield	Strobilanthus Dyeranus	Burma
Black Pepper	Piper Nigrum	India/Ceylon
Torch Ginger	Etlingera Elatior	Celebes/Java
Bleeding Heart Vine	Clerodendrum Thomsoniae	Tropical Africa
Torch Ginger	Etlengera Elatior	Celebes/Java
Coffee	Coffea Arabica	Tropical Africa
Brownea	Brownea Macrophylla	Panama/Colombia
Mother-in-Law Tongue	Sansevieria Trifasciata	Zaire
Sandpaper Vine	Petrea Volubilis	W. Indies/Mexico
Trimezia	Trimezia Martinicensis	Tropical America
Kokio Keokeo	Hibiscus Arnottianus	Hawaii

Plants on Trail From Giant Fern Circle through the Palm Vista Trail

COMMON NAME	BOTANICAL NAME	ORIGIN
Shaving Brush Tree	Pseudobombax Ellipticum	Mexico

Coral tree blossom (*Erythrina bidwillii*)

COMMON NAME	BOTANICAL NAME	ORIGIN
Alo Alo/Hibiscus	Hibiscus Rosa Sinensis	Horticultural
Ceriman	Monstera Deliciosa	Mexico/Central America
Birds Nest	Anthurium Hookeri	Guiana
Philodendron	Philodendron Cannifollium	Brazil
Birds Nest Anthurium	Anthurium Jenmanii	Venezuela/Trinidad
Slipper Flower	Pedilanthus Tithymaloides	Tropical America
Calathea	Calathea Roseopicta	Brazil
Window Pane Palm	Reinhardtia Gracilis	Southern Mexico, Central America
Calathea	Calathea Warscewiczii	Costa Rica
Rhapis Palm	Rhapis Excelsa Variegata	Japan
Rhopaloblaste	Rhopaloblaste Augusta	Nicobar Island
Licuala Palm	Licuala Spinosa	Malaya
Livistona	Livistona Rotundifolia	Malaysia
Ardisia	Ardisia Crispa	Japan/China
Giant Tree Ferns	Cyathea Contaminens	Malaysia
Senegal Date Palm	Phoenix Reclinata	Tropical Africa
Thatch Palm	Thrinax Morrisii	Florida/West Indies
Drivi Drivi	Pseudoeranthemum Laxiflorum	Fiji
Miracle Fruit	Synsepalum Dulcificum	West Africa
Dwarf Ixcra	Ixora Sp.	Horticultural
Jelly Palm	Butia Capitata	Brazil to Uruguay
Orange Costus	Costus Igneus	Brazil
Ming Aralia	Polyscias Fruticosa	India to Polynesia
Oyster Plant	Rhoeo Spathacea	Mexico/Guatemala
Areca Palm	Areca Vestiaria	Celebes
Wanga Palm	Pigafetta Filaris	Malaysia
Medinilla	Medinilla Magnifica	Philippine Islands
Wheel of Fire Tree	Stenocarpus Sinuatus	Australia/Malaysia
Caribbean Royal Palm	Roystonea Oleracea	Caribbean
Black Palm	Normanbya Normanbyi	Australia
Bougainvillea	Bougainvillea Sp.	Brazil
Cabadae Palm	Chrysalidocarpus Cabadae	Extinct in Wild
Sugar Palm	Arenga Pinnata	Malaya
Devil Palm	Aiphanes Caryotifolia	N. South America
Chenille Plant	Acalypha Hispida	East Indies
Yellow Shrimp Plant	Pachstachys Lutea	Peru
Calabash Tree	Crescentia Cujete	Tropical America
Jaboticaba	Myrciaria Cauliflora	Brazil
Heliconia	Heliconia Schiedeana	Mexico
Heliconia	Heliconia Vaginalis	Central American
Heliconia	Heliconia Psittacorum c.v. Sassy	Uncertain
Sausage Tree	Kigelia Pinnata	Tropical America
Jack Fruit	Artocarpus Heterophyllus	S. India/Malaysia
Queensland Umbrella Tree	Brassaia Actinophylla	Australia
Gardenia	Gardenia Jasminoides	China
Heliconia	Heliconia Psittacorum c.v. Lizette	Uncertain
Heliconia	Heliconia Collinsiana	Mexico to Nicaragua
Yellow Ginger	Hedychium Flavescens	India
Glory Bower	Clerodendrum Splendens	Tropical Africa

Plants on the Palm Jungle Trail

COMMON NAME	BOTANICAL NAME	ORIGIN
Alexandra Palm	Archontophoenix Alexandrae	Australia
Caladium	Caladium Lindenii	Colombia
Mountain Apple	Syzygium Malaccense	Malay Peninsula
Little Club Moss	Selaginella Sp.	
Velvet Leaf	Miconia Calvescens	Tropical America

Plants Along Heliconia Trail

COMMON NAME	BOTANICAL NAME	ORIGIN
Ice Cream Banana	Musa X Paradisiaca	Horticultural
Flowering Banana	Musa Ornata	Burma
Heliconia	Heliconia Mariae	Guatemala to Colombia
Heliconia	Heliconia Vellerigera	Colombia to Ecuador/Peru
Blood Leaf Bananna	Musa Sumatrana c.v. Rubra	Sumatra
Heliconia	Heliconia Curtispatha	Nicaragua to Ecuador
Heliconia Yellow Christmas	Heliconia Stricta c.v. Bucky	Guyana
Heliconia	Heliconia Angusta	Brazil
Heliconia	Heliconia Bourgaeana	Mexico
Heliconia	Heliconia Psittacorum c.v. Sherbert	Guyana
Sexy Pink Heliconia	Heliconia Chartacea c.v. Sexy Pink	Brazil
Hanging Lobster Claw	Heliconia Rostrata	Peru
Heliconia	Heliconia Pendula	Brazil
Heliconia	Heliconia Beckneri	Costa Rica
Heliconia	Heliconia Lingulata	Peru to Bolivia
Heliconia	Heliconia Stricta c.v. Tagami	Amazon
Fiery Spike	Aphelandra Aurantiaca	Mexico/South America
Clinostigma Plam	Clinostigma Ponapensis	Ponape
Flowering Ginger	Alpinia Purpurata c.v. Jungle Queen	S. Pacific Is.
Heliconia	Heliconia Ramonensis var. Ramonensis	Costa Rica Panama
Sexy Scarlet Heliconia	Heliconia Chartacea c.v. Sexy Scarlet	Guianas & Brazil
Red Ginger	Hedychium Coccineum c.v. Angustifolium	India
Red Lobster Claw	Heliconia Caribaea	West Indies
Yellow Lobster Claw	Heliconia Caribaea	West Indies
Ctenanthe	Ctenanthe Lubbersiana	Brazil
Rainbow Heliconia	Heliconia Wagneriana	Belize
Tahitian Ginger	Alpinia Purpurata c.v. Tahitian	Pacific Is.
Banyan	Ficus Sp.	India
Peperomia	Peperomia Obtusifolia	Tropical America
Heliconia	Heliconia Bihai c.v. Yellow Dancer	St. Vincent
Night Blooming Jasmine	Cestrum Nocturnum	West Indies
Red Passion Flower	Passiflora Coccinea	Venezuela to Bolivia
Heliconia	Heliconia Psittacorum c.v. Blush	Uncertain
Strawberry Guava	Psidium Littorale	Brazil
Heliconia Lobster Claw	Heliconia Bihai c.v. Lobster	Northern South America
Surinam Cherry	Eugenia Uniflora	Tropical America
Pua Kenikeni	Fagraea Berteriana	Pacific Is.
Starfruit	Averrhoa Carambola	Malay Region
Miracle Leaf	Kalanchoe Pinnata	Uncertain
Heliconia	Heliconia Champneiana	Central America
Heliconia	Heliconia Mutisiana	Colombia
Heliconia	Heliconia Bihai c.v. Nappi Yellow	Guyana
Pink Flowering Banana	Musa Ornata	Pakistan to Burma
Golden Torch Heliconia (hybrid)	Heliconia Psittacorum X H. Spathocircinata	Guyana
Variegated Snake Plant	Sansevieria Trifaciata Var. Laurentii	Africa

Plants Along Cook Pine Trail

COMMON NAME	BOTANICAL NAME	ORIGIN
Bottle Palm	Hyophorbe Lagenicaulis	Mascarene Is.
Red Flowering Banana	Musa Coccinea	Indochina
Ti Plant	Cordyline Terminalis	Eastern Asia
Heliconia	Heliconia Latispatha	Mexico to South America
Dwarf Date Palm	Phoenix Roebelenii	Laos
Money Tree	Dracaena Marginata	Madagascar
Rattlesnake Plant	Calathea Insignis	Ecuador
Ivory Crownshaft Palm	Pinanga Cochinchinensis	Thailand
Alexandra Palm	Archontophoenix Alexandrae	Australia
Lady Palm	Rhapis Humilis	Southern China
Heliconia	Heliconia Collinsiana	S. Mexico to Central Nicaragua
Ice Blue Calathea	Calathea Burle Marxll	Brazil
Hanging Lobsterclaw	Heliconia Rostrata	Peru
Wili Wili	Erythrina Sandwicensis	Hawaii
Loulu Palm	Pritchardia Guadichaudii	Moloka'i, HI
Loulu Palm	Pritchardia Schattaueri	Hawaii
Ho Awa	Pittosporum Hosmeri	Hawaii
Floss Silk Tree	Chorisia Speciosa	Brazil
Fern of the Desert	Lysiloma Thornberi	Southwestern U.S.
Persian Shield	Strobilanthes Dyeranus	Burma
Copperleaf Plant	Acalypha Wilkesiana	Pacific Is.
African Iris	Dietes Vegeta	South Africa
Boat Lily	Rhoeo Spathacea	Mexico to Guatemala
Purple Ground Orchid	Spathoglottis Deplanchii	Malay Archipelago
False Heather	Cuphea Hyssopifolia	Mexico to Guatemala
Clumping Areca Vestiaria	Areca Vestiaria	Celebes Is.
Cook Pine	Araucaria Columnaris	New Caledonia, New Hebrides
Avocado	Persea Americana	Tropical America
Red Orchid Tree	Bauhinia Punctata	Tropical Africa
Blacktree Orchid	Sphaeropteris Medullaris	New Zealand
West Indian Tree Fern	Cyathea Arborea	Tropical America
Warabi/Paco	Diplazium Esculentum	Polynesia
Hawaiian Tree Fern	Cibotium Splendens	Hawaii

COMMON NAME	BOTANICAL NAME	ORIGIN
Hawaiian Tree Fern	Cibotium Glaucum	Hawaii
Taro	10 Varieties—Each Identified by Plant Label	
Palapalai Fern	Microlepia Setosa	Hawaii
Hawaiian Rose	Osteomeles Anthyllidifolia	Hawaii
Amau	Sadleria Cyatheoides	Hawaii
Lizard Plant	Alocasia Portei	Philippine Is.
Kokio Ke'oke'o Hibiscus	Arnottianus	Hawaii
Koki'o Ke'oke'o	Hibiscus Arnottianus Var. Punaluuensis	Oahu, HI
Ma'o Hau Hele	Hibiscus Brackenridgei	Hawaii
Koki'o Ula	Hibiscus Kokio	Hawaii
Koki'o Koki'o	Hibiscus Clayi (H. Newhousei)	Kauai, HI
Ohia	Metrosideros Polymorpha	Hawaii

Hidden lily (*Curcuma elata*).

Plants Along Boulder Creek Trail

Heterosphathe Heterophylla
Asterogyne Martiana
Eucharis Grandiflora
Dracaena Deremensis, c.v. Compacta
Licuala Grandis
Calathea Warscewiczii
Didymochilaena Truncatula
Calathea Libbyana
Calathea Lietzii
Calathea Picturata
Chamaedorea Metallica
Psidium Littorale
Selaginella Willdenovii
Hedychium Flavescens
Calathea Gandersii
Clivia Miniata

Plants At Lily Lake

COMMON NAME	BOTANICAL NAME	ORIGIN
Clinostigma	Clinostigma Harlandii	
Spider Lily	Crinum Amabile	Sumatra
Latanier Feulle	Phoenicophorium Borsigianum	Seychelles Is.
Dwarf Date Palm	Phoenix Roebelenii	Laos/Vietnam
Wandering Jew	Zebrina Pendula	Mexico
Spindle Palm	Hyophorbe Verschaffeltii	Mascarene Island
Manilla Palm	Veitchia Merrillii	Philippines
Veitchia	Veitchia Joannis	Fiji
Ti Plant	Cordyline Terminalis	Eastern Asia
Trimezia	Trimezia Martinicensis	Tropical America
African Iris	Dietes Bicolor	South Africa
Paco/Warabi	Diplazium Esculentum	Asia/Polynesia
Split Leaf Philodendron	Philodendron Selloum	Brazil
Triangle Palm	Neodypsis Decaryi	Madagascar
Red Leaf Heliconia	Heliconia Rubra	Mexico
Hibiscus	Hibiscus Rosa Sinensis	Tropical Asia
Impatiens	Impatiens Wallerana	Tanzania, Mozambique
Blue Fern	Polypodium Aureum	Florida to Argentina
Bloodleaf	Iresine Herbstii	South America
Australian Tree Fern	Sphaeropteris Cooperi	Australia
Giant Alocasia	Alocasia Odora	India to China/Phillippines
Croton	Codiaeum Variegatum	Malayysia
Yucca	Yucca Elephantipes	Mexico
Palm	Chrysalidocarpus Sp.	Malay Peninsula Pacific Is.
Peter Buck Ti	Cordyline Terminalis	Horticultural
Papyrus	Cyperus Papyrus	North Tropical Africa
Water Lily	Nymphaea Sp.	Horticultural
Lotus	Nelumbo Sp.	Asia
Pickerel Weed	Pontederia Cordata	North America
Water Lettuce	Pistia Statiotes	Pantropical
Water Fern	Salvinia Auriculata	Mexico to Argentina
Blue Latan Palm	Latania Loddigesii	Mauritius Is.
Red Sealing Wax Palm	Cytrostachys Renda	Sumatra
Fiji Fan Palm	Pritchardia Pacifica	Fiji
Water Hyacinth	Eichhornia Crassipes	Tropical America
Peacock Hyacinth	Eichhornia Azurea	Brazil
Livistona Palm	Livistona Rotundifolia	Indonesia

COMMON NAME	BOTANICAL NAME	ORIGIN
Travelers Tree	Ravenala Madagascariensis	Madagascar
Black Palm	Normanbya Normanbyi	Australia
Coconut	Cocos Nucifera	World Tropics
Wi Apple	Spondias Cytherea	Society Is.
Chamaedorea	Chamaedorea Cataractarum	Southern Mexico
Ivory Nut Palm	Meteroxylon Vitense	Fiji
Spiny Root Palm	Cryosophila Argentea	Brit.Honduras Guatemala
Queen Palm	Syagrus Romanzoffianum	Brazil to Argentina
Alexandra Palm	Archontophoenix Alexandrae	Australia
Monstera	Monstera Friedrichsthalii	Central America
Mexican Weeping Bamboo	Otatea Aztecorum	Mexico
Ctenanthe	Ctenanthe Lubbersiana	Brazil
Mosaic Fig	Ficus Aspera Var. Parcellii	S. Pacific Is.
Cup of Gold	Solandra Maxima	Mexico
Umbrella Tree	Brassasia Actinophylla	Australia
Bangkok Rose	Mussaenda Erythrophylla	West Africa
Banana	Musa X Paraisiaca	Horticultural
Papaya	Carica Papaya	Tropical America
Australian Flame Tree	Brachychiton Acerifolius	Australia
Mules Foot Fern	Angioteris Evecta	Japan to Australia, Madagascar
Mango	Magnifera Indica	India
Thatch Palm	Coccothrinax Argentea	Hispaniola
Monkeypod	Samanea Saman	Tropical America
Sago Palm	Cycas Revoluta	Japan
Yellow Water Iris	Iris Pseudoacorus	Europe & Africa
Whaleback	Molineria Recurvata	Asia/Australia
Licuala	Licuala Grandis	New Hebrides Is.
Asterogyne	Asterogyne Martiana	Central & South
Spider Lily	Crinum Asiaticum	Tropical Asia
Licuala Palm	Licuala Peltata	India/Burma
Bromeliad	Aechmea Mexicana	Mexico
Lizard Plant	Alocasia Portei	Philippine Is.
African Tulip Tree	Spathodea Campanulata	Africa
Ming Aralia	Polyscias Fruticosa	India to Polynesia
Clerodendrum	Clerodendrum Quadriloculare	New Guinea, Phillippines
Macarthur Palm	Ptychosperma Macarthurii	New Guinea
Christmas Heliconia	Heliconia Angusta	Brazil
Hanging Lobster Claw	Heliconia Rostrata	Peru
Hawaiian Tree Fern	Cibotium Sp.	Hawaii
Cardboard Plant	Zamia Pumila	Florida to Mexico
Pink Quill	Tillandsia Cyanea	Ecuador
Brentwood Tree Fern	Sphaeropteris Cooperi c.v. Brentwood	Australia
Dwarf Papyrus	Cyperus Isocladus	Africa
Red Flowering Banana	Musa Coccinea	Indochina
Bottle Palm	Hyophorbe Lagenicaulis	Mascarene Is.
Baby Doll Ti	Cordyline Terminalis c.v. Baby Doll	Horticultural
Betel Nut Palm	Areca Catechu	Malay Peninsula
Money Tree	Dracaena Marginata	Madagascar
Ti Plant	Cordyline Terminalis	Eastern Asia
Queen Sago	Cycas Circinalis	Old World Tropics
Solitaire Palm	Ptychosperma Elegans	Australia
Corn Plant	Dracaena Fragrans	Upper Guinea
Chinese Fan Palm	Livistona Chinensis	Japan/China

COMMON NAME	BOTANICAL NAME	ORIGIN
Pink Flowering Banana	Musa Velutina	Assam
Heliconia	Heliconia Tortuosa	Costa Rica
Pinanga Palm	Pinanga Kuhlii	Java, Sumatra
Dumb Cane	Dieffenbachia Maculata	Central & South America
Basselinia	Basselinia Eriostachys	Trinidad/Guyana
Flamingo Plant	Justicia Carnea	South America
Wanga Palm	Pigafetta Filaris	Celebes Moluccas
Gaussia Palm	Gaussai Attenuata	Puerto Rico
Royal Palm	Roystonea Sp.	
Pink Ginger	Alpinia Purpurata	Pacific Is.
Blue Trumpet Vine	Thunbergia Grandiflora	India
Heliconia	Heliconia Episcopalis	Amazon
Hong Kong Orchid Tree	Bauhinia Blakeana	China
Heliconia	Heliconia Indica var. Striata	S. Pacific Is.
Heliconia	Heliconia Latispatha	Mexico to S. America
Common Apé	Xanthosoma Robustum	Mexico
Wild Poinsettia	Warszewiczia Coccinea	Trinidad to Brazil
Fern Leaf Aralia	Polyscias Filicifolia	Pacific Is.
Soft Tip Agave	Agave Attenuata	Mexico
Laua'e Fern	Polypoduim Scolopendria	Old World Tropics
Banyan	Ficus Sp.	India

Red mountain apple tree blossoms (*Syzygium malaccense*).

How to Visit
Hawaii Tropical Botanical Garden

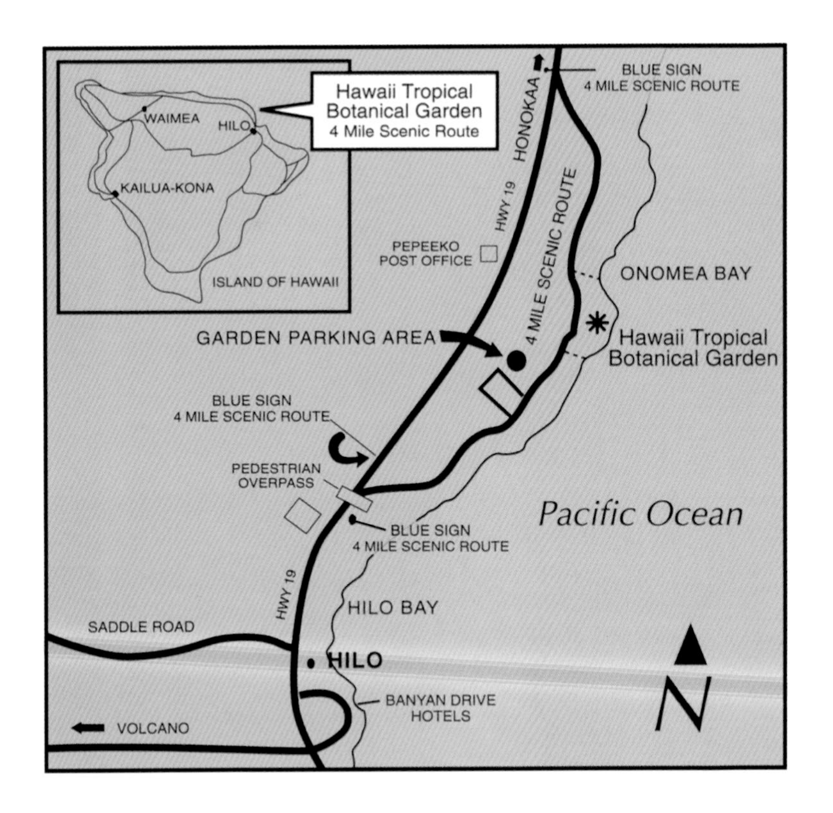

Getting to the Garden

FROM HILO: Travel north on Hwy. 19 along the Hamakua Coast for approximately seven miles. Watch for the blue signs which say, "Scenic Route 4 Miles Long". Turn right onto the Scenic Route and drive about two miles to the Garden's Visitor Center.

FROM KONA: Drive to Waimea and then follow Hwy. 19 south towards Hilo. About five miles after the towns of Hakalau and Honomu, watch for a blue sign on the right saying "Scenic Route 4 Miles Long". Turn left at intersection. Continue about two miles to the Garden's Visitor Center.

Garden Hours

The Garden is open to visitors from 9:00 a.m. to 5:00 p.m. (last entry into the Garden at 4:00 p.m.) seven days a week. The Garden is closed on Thanksgiving Day, Christmas Day and New Year's Day. All visitors are provided with trail maps for their own self-guided tour. The Garden trails are paved enabling persons in wheelchairs to tour approximately ninety percent of the Garden. Persons in wheelchairs are provided with a Golf Cart ride into and out of the Garden for accessibility.

Admission Fees

Daily admission is charged to visit the Garden. There are also various levels of memberships available. All admissions, memberships and donations are tax deductible to the full extent allowed by law.

What if it Rains?

Don't worry- the Garden is beautiful in the rain! Free use of umbrellas are provided at the Garden's Visitor Center. Hundreds of visitors enjoy the rare experience of walking through a tropical rainforest in the rain.

Garden Gift Shop and Museum

Located at the Visitor Center the Garden's elegant Gift Shop offers a fine array of art glass, ceramics, jewelry and other items, handcrafted by local artisans and all personally selected by Pauline Lutkenhouse to ensure the highest quality. Custom designed tee shirts on a botanical theme are available in a wide assortment and there is a broad selection of books on tropical plants and Hawaii.

Garden Awards

In 1993 the Hawaii Tropical Botanical Garden received several prestigious Awards for excellence. One of these was the Kahili Award for the "Best Attraction" in the state of Hawaii. The Garden was cited for strengthening the aloha spirit by offering an attraction that "educates, preserves, and promotes unique qualities of our island". In addition, the Garden received the Richard Smart Community Achievement Award for providing free visits to schoolchildren; for service to the Boys and Girls Clubs and the Kamehameha Canoe Club; and for offering teaching programs to help visitors understand the island's environment, culture and history .

ABOVE AND RIGHT: PAULINE LUTKENHOUSE AND THE GARDEN BOOK AND GIFT SHOP WHICH SHE ESTABLISHED AND MANAGES. YOU WILL FIND WONDERFUL MEMENTOS OF YOUR VISIT TO THE GARDEN.